BRIGHT AND SHINY

Venus is sometimes called the "Evening Star" or the "Morning Star" because it shines so brightly in the sky, especially around sunrise and sunset.

MEASURE OF DISTANCE

A light-year is the distance that light travels in one year is about 9.5 trillion kilometers!

Thank you for purchasing this book!

We would highly value your feedback if you could consider leaving a review at the point of purchase after enjoying this picture book. Your input holds significant meaning for us and assists potential readers in making well-informed choices.

STARS GALORE

The night sky is like a glittering blanket, filled with stars. In fact, there are so many stars that scientists estimate there are billions of them just in our own galaxy, the Milky Way!

LIGHT TRAVEL

Some stars are incredibly far away from us, so far that it takes millions or even billions of years for their light to reach Earth.

TWINKLING STARS

Stars appear to twinkle in the night sky because of the way the Earth's atmosphere bends and moves the light coming from them.

GUIDING LIGHT

The North Star, also called Polaris, is special because it's very close to the North Celestial Pole. It's like a compass in the sky, helping travelers find their way when they're lost.

GALACTIC BAND

The Milky Way is not just a candy bar! It's actually the name of our galaxy, and it looks like a faint band of light stretching across the sky on clear nights.

OUR MOON

The Moon is Earth's natural satellite, and it's the fifth largest moon in our solar system.

MOONLIGHT MAGIC

The Moon doesn't make its own light; it reflects light from the Sun, which is why it shines so brightly in the night sky.

PHASES OF THE MOON

The Moon goes through different phases during a month, from a thin crescent to a full circle and back again. These phases happen because of the way the Moon moves around Earth and how the Sun's light hits it.

ECLIPSES

There are two types of eclipses: lunar and solar. A lunar eclipse happens when Earth gets between the Sun and the Moon, casting a shadow on the Moon. A solar eclipse occurs when the Moon passes between the Sun and Earth, blocking the Sun's light.

SHOOTING STARS

Those streaks of light in the sky are called shooting stars or meteors. They're caused by small rocks burning up as they enter Earth's atmosphere.

METEOR SHOWERS

Sometimes, there are lots of shooting stars all at once, and that's called a meteor shower. The Perseid meteor shower in August is one of the most famous.

NEBULAE NURSERIES

Nebulae are huge clouds of gas and dust in space where new stars are born. The Orion Nebula is one of the most famous ones.

STORMY PLANET

Jupiter, the largest planet in our solar system, has a giant storm called the Great Red Spot that's been swirling for centuries.

RINGED BEAUTY

Saturn is famous for its beautiful rings made of ice and rock particles.

BRIGHT AND SHINY

Venus is sometimes called the "Evening Star" or the "Morning Star" because it shines so brightly in the sky, especially around sunrise and sunset.

RED NEIGHBOR

Mars is known as the "Red Planet" because of its rusty color, caused by iron-rich minerals in its soil.

SPACE STATION IN ORBIT

The International Space Station (ISS) is like a science laboratory floating in space. Sometimes you can see it passing overhead like a bright star!

SUNRISES AND SUNSETS GALORE

Astronauts aboard the ISS experience about 16 sunrises and sunsets every day because they orbit the Earth so quickly.

CLOSEST NEIGHBOR

The Andromeda Galaxy is the nearest spiral galaxy to our Milky Way. On a clear night, you can see it as a fuzzy patch of light.

HUBBLE'S VIEW

The Hubble Space Telescope has taken some of the most breathtaking pictures of distant galaxies, stars, and nebulae.

MYSTERIOUS BLACK HOLES

Black holes are super dense objects with incredibly strong gravity. They're so powerful that not even light can escape from them!

CONSTELLATIONS

These are groups of stars that form recognizable patterns. For example, Ursa Major contains the Big Dipper and is also known as the Great Bear.

HUNTER'S CONSTELLATION

Orion is named after a famous hunter from Greek mythology. It contains bright stars like Betelgeuse and Rigel.

MEASURE OF DISTANCE

A light-year is the distance that light travels in one year, which is about 9.5 trillion kilometers!

OUR SHINING STAR

The Sun is a star, but it's special to us because it's at the center of our solar system. It provides light and heat to Earth, making life possible.

SUN'S SIZE

The Sun is so big that about 1.3 million Earths could fit inside it!

SUN'S TEMPERATURE

The surface of the Sun is incredibly hot, with temperatures reaching about 5,500 degrees Celsius (9,932 degrees Fahrenheit).

ASTRONOMY

The study of stars, planets, and other celestial objects is called astronomy, and scientists who study them are called astronomers.

ENDLESS EXPLORATION

The universe is vast and full of wonders waiting to be discovered. Who knows what else is out there beyond our night sky? Exploring space is an exciting adventure that continues to captivate scientists and dreamers alike!

FAMOUS SHAPES

The Big Dipper, also known as the Plough, is one of the most easily recognizable star patterns in the sky. It looks like a big spoon or ladle.

www.ingramcontent.com/pod-product-compliance
Lightning Source LLC
Chambersburg PA
CBHW060804290526
45792CB00005BA/1523